HOW COSMIC FORCES SHAPE OUR DESTINIES

AND

TALKING WITH THE PLANETS

HOW COSMIC FORCES SHAPE OUR DESTINIES

AND

TALKING WITH THE PLANETS

NIKOLA TESLA

WILDSIDE PRESS

CONTENTS

HOW COSMIC FORCES SHAPE OUR DESTINIES 5

TALKING WITH THE PLANETS 13

"How Cosmic Forces Shape Our Destinies" was originally published in *New York American*, February 7, 1915.

"Talking with the Planets" was originally published in *Collier's Weekly*, February 19, 1901.

Published by Wildside Press LLC.
wildsidepress.com

HOW COSMIC FORCES
SHAPE OUR DESTINIES

Every living being is an engine geared to the wheelwork of the universe. Though seemingly affected only by its immediate surrounding, the sphere of external influence extends to infinite distance. There is no constellation or nebula, no sun or planet, in all the depths of limitless space, no passing wanderer of the starry heavens, that does not exercise some control over its destiny—not in the vague and delusive sense of astrology, but in the rigid and positive meaning of physical science.

More than this can be said. There is no thing endowed with life—from man, who is enslaving the elements, to the humblest creature—in all this world that does not sway it in turn. Whenever action is born from force, though it be infinitesimal, the cosmic balance is upset and universal motion result.

Herbert Spencer has interpreted life as a continuous adjustment to the environment, a definition of this inconceivably complex manifestation quite in accord with advanced scientific thought, but, perhaps, not broad enough to express our present views. With each step forward in the investigation of its laws and mysteries our conceptions of nature and its phases have been gaining in depth and breadth.

In the early stages of intellectual development man was conscious of but a small part of the macrocosm. He knew nothing of the wonders of the microscopic world, of the molecules composing it, of the atoms making up the molecules and of the dwindlingly small world of electrons within the atoms. To him life was synonymous with voluntary motion and action. A plant did not suggest to him what it does to us—that it lives and feels, fights for its existence, that it suffers and enjoys. Not only have we found this to be true, but we have ascertained that even matter called inorganic, believed to be dead, responds to irritants and gives unmistakable evidence of the presence of a living principle within.

Thus, everything that exists, organic or inorganic, animated or inert, is susceptible to stimulus from the outside. There is no gap between, no break of continuity, no special and distinguishing vital agent. The same law governs all matter, all the universe is alive. The momentous question of Spencer, "What is it that causes inorganic matter to run into organic forms!" has been answered. It is the sun's heat and light. Wherever they are there is life. Only in the boundless wastes of interstellar space, in the eternal darkness and cold, is animation suspended, and, possibly, at the temperature of absolute zero all matter may die.

MAN AS A MACHINE

This realistic aspect of the perceptible universe, as a clockwork wound up and running down, dispensing with the necessity of a hypermechanical vital principle, need not be in discord with our religious and artistic aspirations—those undefinable and beautiful efforts through which the human mind endeavors to free itself from material bonds. On the contrary, the better understanding of nature, the consciousness that our knowledge is true, can only be all the more elevating and inspiring.

It was Descartes, the great French philosopher, who in the seventeenth century, laid the first foundation to the mechanistic theory of life, not a little assisted by Harvey's epochal discovery of blood circulation. He held that animals were simply automata without consciousness and recognized that man, though possessed of a higher and distinctive quality, is incapable of action other than those characteristic of a machine. He also made the first attempt to explain the physical mechanism of memory. But in this time many functions of the human body were not as yet understood, and in this respect some of his assumptions were erroneous.

Great strides have since been made in the art of anatomy, physiology and all branches of science, and the workings of the man-machine are now perfectly clear. Yet the very fewest among us are able to trace their actions to primary external causes. It is indispensable to the arguments I shall advance to keep in mind the main facts which I have myself established in years of close reasoning and observation and which may be summed up as follows:

1. The human being is a self-propelled automaton entirely under the control of external influences. Willful and predetermined though they appear, his actions are governed not from within, but from without. He is like a float tossed about by the waves of a turbulent sea.

2. There is no memory or retentive faculty based on lasting impression. What we designate as memory is but increased responsiveness to repeated stimuli.

3. It is not true, as Descartes taught, that the brain is an accumulator. There is no permanent record in the brain, there is no stored knowledge. Knowledge is something akin to an echo that needs a disturbance to be called into being.

4. All knowledge or form conception is evoked through the medium of the eye, either in response to disturbances directly received on the retina or to their fainter secondary effects and reverberations. Other sense organs can only call forth feelings which have no reality of existence and of which no conception can be formed

5. Contrary to the most important tenet of Cartesian philosophy that the perceptions of the mind are illusionary, the eye transmits to it the true and accurate likeness of external things. This is because light propagates in straight lines and the image cast on the retina is an exact reproduction of the external form and one which, owing to the mechanism of the optic nerve, can not be distorted in the transmission to the brain. What is more, the process must be reversible, that in to say, a form brought to consciousness can, by reflex action, reproduce the original image on the retina just as an echo can reproduce the original disturbance If this view is borne out by experiment an immense revolution in all human relations and departments of activity will be the consequence.

NATURAL FORCES INFLUENCE US

Accepting all this as true let us consider some of the forces and influences which act on such a wonderfully complex automatic engine with organs inconceivably sensitive and delicate, as it is carried by the spinning terrestrial globe in lightning flight through space.

For the sake of simplicity we may assume that the earth's axis is perpendicular to the ecliptic and that the human automaton is at the equator. Let his weight be one hundred and sixty pounds then, at the rotational velocity of about 1,520 feet per second with which he is whirled around, the mechanical energy stored in his body will be nearly 5,780,000 foot pounds, which is about the energy of a hundred-pound cannon ball.

This momentum is constant as well as upward centrifugal push, amounting to about fifty-five hundredth of a pound, and both will probably be without marked influence on his life functions. The sun, having a mass 332,000 times that of the earth, but being 23,000 times farther, will attract the automaton with a force of about one-tenth of one pound, alternately increasing and diminishing his normal weight by that amount

Though not conscious of these periodic changes, he is surely affected by them.

The earth in its rotation around the sun carries him with the prodigious speed of nineteen miles per second and the mechanical energy imparted to him is over 25,160,000,000 foot pounds. The largest gun ever made in Germany hurls a projectile weighing one ton with a muzzle velocity of 3,700 feet per second, the energy being 429,000,000 foot pounds. Hence the momentum of the automaton's body is nearly sixty times greater. It would be sufficient to develop 762,400 horse-power for one minute, and if the motion were suddenly arrested the body would be instantly exploded with a force sufficient to carry a projectile weighing over sixty tons to a distance of twenty-eight miles.

This enormous energy is, however, not constant, but varies with the position of the automaton in relation to the sun. The circumference of the earth has a speed of 1,520 feet per second, which is either added to or subtracted from the translatory velocity of nineteen miles through space. Owing to this the energy will vary from twelve to twelve hours by an amount approximately equal to 1,533,000,000 foot pounds, which means that energy streams in some unknown way into and out of the body of the automaton at the rate of about sixty-four horse-power.

But this is not all. The whole solar system is urged towards the remote constellation Hercules at a speed which some estimate at some twenty miles per second and owing to this there should be similar annual changes in the flux of energy, which may reach the appalling figure of over one hundred billion foot pounds. All these varying and purely mechanical effects are rendered more complex through the inclination of the orbital planes and many other permanent or casual mass actions.

This automaton, is, however subjected to other forces and influences. His body is at the electric potential of two billion volts, which fluctuates violently and incessantly. The whole earth is alive with electrical vibrations in which he takes part. The atmosphere crushes him with a pressure of from sixteen to twenty tons, according, to barometric condition. He receives the energy of the sun's rays in varying intervals at a mean rate of about forty foot pounds per second, and is subjected to periodic bombardment of the sun's particles, which pass through his body as if it were tissue paper. The air is rent with sounds which beat on his eardrums, and he is shaken by the unceasing tremors of the earth's crust. He is exposed to great temperature changes, to rain and wind.

What wonder then that in such a terrible turmoil, in which cast iron existence would seem impossible, this delicate human engine should act in an exceptional manner? If all automata were in every respect alike they would react in exactly the same way, but this is not the case. There is concordance in response to those disturbances only which are most frequently repeated, not to all. It is quite easy to provide two electrical systems which, when subjected to the same influence, will behave in just the opposite way.

So also two human beings, and what is true of individuals also holds good for their large aggregations. We all sleep periodically. This is not an indispensable physiological necessity any more than stoppage at intervals is a requirement for an engine. It is merely a condition gradually imposed upon us by the diurnal revolution of the globe, and this is one of the many evidences of the truth of the mechanistic theory. We note a rhythm or ebb and tide, in ideas and opinions, in financial and political movements, in every department of our intellectual activity.

HOW WARS ARE STARTED

It only shows that in all this a physical system of mass inertia is involved which affords a further striking proof. If we accept the theory as a fundamental truth and, furthermore, extend the limits of our sense perceptions beyond those within which we become conscious of the external impressions, then all the states in human life, however unusual, can be plausibly explained. A few examples may be given in illustration.

The eye responds only to light vibrations through a certain rather narrow range, but the limits are not sharply defined. It is also affected by vibrations beyond, only in lesser degree. A person may thus become aware of the presence of another in darkness, or through intervening obstacles, and people laboring under illusions ascribe this to telepathy. Such transmission of thought is absurdly impossible.

The trained observer notes without difficulty that these phenomena are due to suggestion or coincidence. The same may be said of oral impressions, to which musical and imitative people are especially susceptible. A person possessing these qualities will often respond to mechanical shocks or vibrations which are inaudible.

To mention another instance of momentary interest reference may be made to dancing, which comprises certain harmonious muscular contractions and contortions of the body in response to a rhythm. How they come to be in vogue just now, can be satisfactorily explained by supposing the existence of some new periodic disturbances in the environment, which are transmitted through the air or the ground and may be of mechanical, electrical or other character.

Exactly so it is with wars, revolutions and similar exceptional states of society.

Though it may seem so, a war can never be caused by arbitrary acts of man.

It is invariably the more or less direct result of cosmic disturbance in which the sun is chiefly concerned.

In many international conflicts of historical record which were precipitated by famine, pestilence or terrestrial catastrophes the direct dependence of the sun is unmistakable. But in most cases the underlying primary causes are numerous and hard to trace.

In the present war it would be particularly difficult to show that the apparently willful acts of a few individuals were not causative.

Be it so, the mechanistic theory, being founded on truth demonstrated in everyday experience, absolutely precludes the possibility of such a state being anything but the inevitable consequence of cosmic disturbance.

The question naturally presents itself as to whether there is some intimate relation between wars and terrestrial upheavals. The latter are of decided influence on temperament and disposition, and might at times be instrumental in accelerating the clash but aside from this there seems to be no mutual dependence, though both may be due to the same primary cause.

What can be asserted with perfect confidence is that the earth may be thrown into convulsions through mechanical effects such as are produced in modern warfare. This statement may be startling, but it admits of a simple explanation.

Earthquakes are principally due to two causes—subterranean explosions or structural adjustments. The former are called volcanic, involve immense energy and are hard to start. The latter are named tectonic; their energy is comparatively insignificant and they can be caused by the slightest shock or tremor. The frequent slides in the Culebra are displacements of this kind.

WAR AND THE EARTHQUAKE

Theoretically, it may be said that one might think of a tectonic earthquake and cause it to occur as a result of the thought, for just preceding the release the mass may be in the most delicate balance. There is a popular error in regard to the energy of such displacements. In a case recently reported as quite extraordinary, extending as it did over a vast territory, the energy was estimated at 65,000,000,000,000 foot tons. Assuming even that the whole work was performed in one minute it would only be equivalent to that of 7,500,000 horse-power during one year, which seems much, but is little for a terrestrial upheaval. The energy of the sun's rays falling on the same area is a thousand times greater.

The explosions of mines, torpedoes, mortars and guns develop reactive forces on the ground which are measured in hundreds or even thousands of tons and make themselves felt all over the globe.

Their effect, however, may be enormously magnified by resonance. The earth is a sphere of a rigidity slightly greater than that of steel and vibrates once in about one hour and forty-nine minutes.

If, as might well be possible, the concussions happen to be properly timed their combined action could start tectonic adjustments in any part of the earth, and the Italian calamity may thus have been the result of explosions in France. That man can produce such terrestrial convulsions is beyond any doubt, and the time may be near when it will be done for purposes good or apt.

TALKING WITH THE PLANETS

Editor's Note.—Mr. Nikola Tesla has accomplished some marvellous results in electrical discoveries. Now, with the dawn of the new century, he announces an achievement which will amaze the entire universe, and which eclipses the wildest dream of the most visionary scientist. He has received communication, he asserts, from out the great void of space: a call from the inhabitants of Mars, or Venus, or some other sister planet! And, furthermore, noted scientists like Sir Norman Lockyer are disposed to agree with Mr. Tesla in his startling deductions.

Mr. Tesla has not only discovered many important principles, but most of his inventions are in practical use: notably in the harnessing of the Titanic forces of Niagara Falls, and the discovery of a new light by means of a vacuum tube. He has, he declares, solved the problem of telegraphing without wires or artificial conductors of any sort, using the earth as his medium. By means of this principle he expects to be able to send messages under the ocean, and to any distance on the earth's surface. Interplanetary communication has interested him for years, and he sees no reason why we should not soon be within talking distance of Mars or of all worlds in the solar system that may be tenanted by intelligent beings.

At the request of COLLIER'S WEEKLY, Mr. Tesla presents herewith a frank statement of what he expects to accomplish and how he hopes to establish communication with the planets.

EXPERIMENTS IN COLORADO

The idea of communicating with the inhabitants of other worlds is an old one. But for ages it has been regarded merely as a poet's dream, forever unrealizable. And with the invention and perfection of the telescope and the ever-widening knowledge of the heavens, its hold upon our imaginations has been increased, and the scientific achievements during the latter part of the nineteenth century, together with the development of the tendency toward the nature ideal of Goethe, have intensified it to such a degree that it seems as if it were destined to become the dominating idea of the century that has just begun. The desire to know something of our neighbors in the immense depths of space does not spring from idle curiosity nor from thirst for knowledge, but from a deeper cause, and it is a feeling firmly rooted in the heart of every human being capable of thinking at all

Whence, then, does it come? Who knows? Who can assign limits to the subtlety of nature's influences? Perhaps, if we could clearly perceive all the intricate mechanism of the glorious spectacle that is continually unfolding before us, and could, also, trace this desire to its distant origin, we might find it in the sorrowful vibrations of the earth which began when it parted from its celestial parent.

But in this age of reason it is not astonishing to find persons who scoff at the very thought of effecting communication with a planet. First of all, the argument is made that there is only a small probability of other planets being inhabited at all. This argument has never appealed to me. In the solar system, there seem to be only two planets—Venus and Mars—capable of sustaining life such as ours: but this does not mean that there might not be on all of them some other forms of life. Chemical processes may be maintained without the aid of oxygen, and it is still a question whether chemical processes are absolutely necessary for the sustenance of organized beings. My idea is that the development of life must lead to forms of existence that will be possible without nourishment and which will not be shackled by consequent limitations. Why should a living being not be able to obtain all the energy it needs for the performance of its life functions from the environment, instead of through consumption of food, and transforming, by a complicated process, the energy of chemical combinations into life-sustaining energy?

If there were such beings on one of the planets we should know next to nothing about them. Nor is it necessary to go so far in our assumptions, for we can readily conceive that, in the same degree as the atmosphere diminishes in density, moisture disappears and the planet freezes up, organic life might also undergo corresponding modifications, leading finally to forms which, according to our present ideas of life, are impossible. I will readily admit, of course, that if there should be a sudden catastrophe of any kind all life processes might be arrested; but if the change, no matter how great, should be gradual, and occupied ages, so that the ultimate results could be intelligently foreseen, I cannot but think that reasoning beings would still find means of existence. They would adapt themselves to their constantly changing environment. So I think it quite possible that in a frozen planet, such as our moon is supposed to be, intelligent beings may still dwell, in its interior, if not on its surface.

SIGNALLING AT 100,000,000 MILES!

Then it is contended that it is beyond human power and ingenuity to convey signals to the almost inconceivable distances of fifty million or one hundred million miles. This might have been a valid argument formerly. It is not so now. Most of those who are enthusiastic upon the subject of interplanetary communication have reposed their faith in the light-ray as the best possible medium of such communication. True, waves of light, owing to their immense rapidity of succession, can penetrate space more readily than waves less rapid, but a simple consideration will show that by their means an exchange of signals between this earth and its companions in the solar system is, at least now, impossible. By way of illustration, let us suppose that a square mile of the earth's surface—the smallest area that might possibly be within reach of the best telescopic vision of other worlds— were covered with incandescent lamps, packed closely together so as to form, when illuminated, a continuous sheet of light. It would require not less than one hundred million horse-power to light this area of lamps, and this is many times the amount of motive power now in the service of man throughout the world.

But with the novel means, proposed by myself, I can readily demonstrate that, with an expenditure not exceeding two thousand horse-power, signals can be transmitted to a planet such as Mars with as much exactness and certitude as we now send messages by wire from New York to Philadelphia. These means are the result of long-continued experiment and gradual improvement.

Some ten years ago, I recognized the fact that to convey electric currents to a distance it was not at all necessary to employ a return wire, but that any amount of energy might be transmitted by using a single wire. I illustrated this principle by numerous experiments, which, at that time, excited considerable attention among scientific men.

This being practically demonstrated, my next step was to use the earth itself as the medium for conducting the currents, thus dispensing with wires and all other artificial conductors. So I was led to the development of a system of energy transmission and of telegraphy without the use of wires, which I described in 1893. The difficulties I encountered at first in the transmission of currents through the earth were very great. At that time I had at hand only ordinary apparatus, which I found to be ineffective, and I concentrated my attention immediately upon perfecting machines for this special purpose. This work consumed a number of years, but I finally vanquished all difficulties and succeeded in producing a machine which, to explain its operation in plain language, resembled a pump in its action, drawing electricity from the earth and driving it back into the same at an enormous rate, thus creating ripples or disturbances which, spreading through the earth as through a wire, could be detected at great distances by carefully attuned receiving circuits. In this manner I was able to transmit to a distance, not only feeble effects for the purposes of signalling, but considerable amounts of energy, and later discoveries I made convinced me that I shall ultimately succeed in conveying power without wires, for industrial purposes, with high economy, and to any distance, however great.

EXPERIMENTS IN COLORADO

To develop these inventions further, I went to Colorado in where I continued my investigations along these and other lines, one of which in particular I now consider of even greater importance than the transmission of power without wires. I constructed a laboratory in the neighborhood of Pike's Peak. The conditions in the pure air of the Colorado Mountains proved extremely favorable for my experiments, and the results were most gratifying to me. I found that I could not only accomplish more work, physically and mentally, than I could in New York, but that electrical effects and changes were more readily and distinctly perceived. A few years ago it was virtually impossible to produce electrical sparks twenty or thirty foot long; but I produced some more than one hundred feet in length, and this without difficulty. The rates of electrical movement involved in strong induction apparatus had measured but a few hundred horse-power, and I produced electrical movements of rates of one hundred and ten thousand horse-power. Prior to this, only insignificant electrical pressures were obtained, while I have reached fifty million volts.

The accompanying illustrations, with their descriptive titles, taken from an article I wrote for the "Century Magazine," may serve to convey an idea of the results I obtained in the directions indicated.

Many persons in my own profession have wondered at them and have asked what I am trying to do. But the time is not far away now when the practical results of my labors will be placed before the world and their influence felt everywhere. One of the immediate consequences will be the transmission of messages without wires, over sea or land, to an immense distance. I have already demonstrated, by crucial tests, the practicability of signalling by my system from one to any other point of the globe, no matter how remote, and I shall soon convert the disbelievers.

I have every reason for congratulating myself that throughout these experiments, many of which were exceedingly delicate and hazardous, neither myself nor any of my assistants received any injury. When working with these powerful electrical oscillations the most extraordinary phenomena take place at times. Owing to some interference of the oscillations, veritable balls of fire are apt to leap out to a great distance, and if any one were within or near their paths, he would be instantly destroyed. A machine such as I have used could easily kill, in an instant, three hundred thousand persons. I observed that the strain upon my assistants was telling, and some of them could not endure the extreme tension of the nerves. But these perils are now entirely overcome, and the operation of such apparatus, however powerful, involves no risk whatever.

As I was improving my machines for the production of intense electrical actions, I was also perfecting the means for observing feeble effects. One of the most interesting results, and also one of great practical importance, was the development of certain contrivances for indicating at a distance of many hundred miles an approaching storm, its direction, speed and distance travelled. These appliances are likely to be valuable in future meteorological observations and surveying, and will lend themselves particularly to many naval uses.

It was in carrying on this work that for the first time I discovered those mysterious effects which have elicited such unusual interest. I had perfected the apparatus referred to so far that from my laboratory in the Colorado mountains I could feel the pulse of the globe, as it were, noting every electrical change that occurred within a radius of eleven hundred miles.

TERRIFIED BY SUCCESS

I can never forget the first sensations I experienced when it dawned upon me that I had observed something possibly of incalculable consequences to mankind. I felt as though I were present at the birth of a new knowledge or the revelation of a great truth. Even now, at times, I can vividly recall the incident, and see my apparatus as though it were actually before me. My first observations positively terrified me, as there was present in them something mysterious, not to say supernatural, and I was alone in my laboratory at night; but at that time the idea of these disturbances being intelligently controlled signals did not yet present itself to me.

The changes I noted were taking place periodically, and with such a clear suggestion of number and order that they were not traceable to any cause then known to me. I was familiar, of course, with such electrical disturbances as are produced by the sun, Aurora Borealis and earth currents, and I was as sure as I could be of any fact that these variations were due to none of these causes. The nature of my experiments precluded the possibility of the changes being produced by atmospheric disturbances, as has been rashly asserted by some. It was some time afterward when the thought flashed upon my mind that the disturbances I had observed might be due to an intelligent control. Although I could not decipher their meaning, it was impossible for me to think of them as having been entirely accidental. The feeling is constantly growing on me that I had been the first to hear the greeting of one planet to another. A purpose was behind these electrical signals; and it was with this conviction that I announced to the Red Cross Society, when it asked me to indicate one of the great possible achievements of the next hundred years, that it would probably be the confirmation and interpretation of this planetary challenge to us.

Since my return to New York more urgent work has consumed all my attention; but I have never ceased to think of those experiences and of the observations made in Colorado. I am constantly endeavoring to improve and perfect my apparatus, and just as soon as practicable I shall again take up the thread of my investigations at the point where I have been forced to lay it down for a time.

COMMUNICATING WITH THE MARTIANS

At the present stage of progress, there would be no insurmountable obstacle in constructing a machine capable of conveying a message to Mars, nor would there be any great difficulty in recording signals transmitted to us by the inhabitants of that planet, if they be skilled electricians. Communication once established, even in the simplest way, as by a mere interchange of numbers, the progress toward more intelligible communication would be rapid. Absolute certitude as to the receipt and interchange of messages would be reached as soon as we could respond with the number "four," say, in reply to the signal "one, two, three." The Martians, or the inhabitants of whatever planet had signalled to us, would understand at once that we had caught their message across the gulf of space and had sent back a response. To convey a knowledge of form by such means is, while very difficult, not impossible, and I have already found a way of doing it.

What a tremendous stir this would make in the world! How soon will it come? For that it will some time be accomplished must be clear to every thoughtful being.

Something, at least, science has gained. But I hope that it will also be demonstrated soon that in my experiments in the West I was not merely beholding a vision, but had caught sight of a great and profound truth.

www.ingramcontent.com/pod-product-compliance
Lightning Source LLC
Chambersburg PA
CBHW012011190326
41520CB00025B/7511